HOARD'S DAIRYMAN

Dairy Collectibles

A pictorial guide to
collecting and identifying items
related to dairying.
Includes a
buyers' price guide.

by James S. Baird

Third edition

P O Box 801
Fort Atkinson WI 53538-0801

Publisher of
HOARD'S DAIRYMAN
The National Dairy Farm Magazine

ISBN 0-923147-24-0

FOREWORD

Some years ago, I began collecting dairy artifacts in a modest fashion. It was out of necessity, however, that in 1971 I became serious about it. That was when I accepted the honor and challenge of designing a dairy exhibit for the Hoard Historical Museum.

The occasion was the 100th anniversary of the founding of the Wisconsin Dairymen's Association and the birth of dairying as a great industry in this state. The museum seemed a natural place to record the early years of dairying. This southeastern part of Wisconsin, with its roots deep in the industry, seemed a fertile area to scour for old artifacts. And it was.

During our committee's search, however, we discovered a problem. To our surprise, we could find no comprehensive publication to guide our collecting. Pamphlets and books abounded on nearly every other collectible, but there were few on the many dairy-related items that existed.

The exhibit was completed in 1972, and since then the National Dairy Shrine has selected the museum site as its permanent home. A large museum complex has been built.

This handbook was assembled as a starting point for identifying dairy artifacts and is, in no way, meant to be an in-depth study. The collectibles shown here are from the museum collection and from a number of private collections.

It is very satisfying to capture history through preserving the actual dairy tools of our ancestors. If this book stimulates an interest in these old treasures, then my purpose will have been accomplished.

J. S.B.

ACKNOWLEDGEMENTS

Many people were of invaluable help in the preparation of this book. Listed here are some of the collectors who furnished help and artifacts for this effort:

Rena Amacher
Hans Andregg
Marcia Baird
Verna Baird
Kathy Beane
Laura Beane
Melvin Bigler
James R. Borcherding
Dr. Paul Dettloff
J. S. Johnson
Hank and Norma Kosharek
Rachael Morris
Bill and Leona Niebler
Beverly Schroeder
Dave and Jean Tyler
Howard Voegeli
Bill and Jan Ward
Dennis Yaeggi

Special thanks to the Hoard Historical Museum and its staff.

Many of the items presented here are on display in the National Dairy Shrine building which is adjacent to the Hoard Historical Museum located in Fort Atkinson, Wisconsin.

The museum complex is open Tuesday through Saturday 9:30 a.m. to 3:30 p.m . and the first Sunday of the month from 1:00 p.m. to 5:00 p.m. except on holidays.

Large groups should call for an appointment.

Museum phone (414) 563-7769.

Thanks also should be extended to the
Cream Separator & Dairy Collectors Association
Rt. 3, Arcadia, WI 54612

CONTENTS

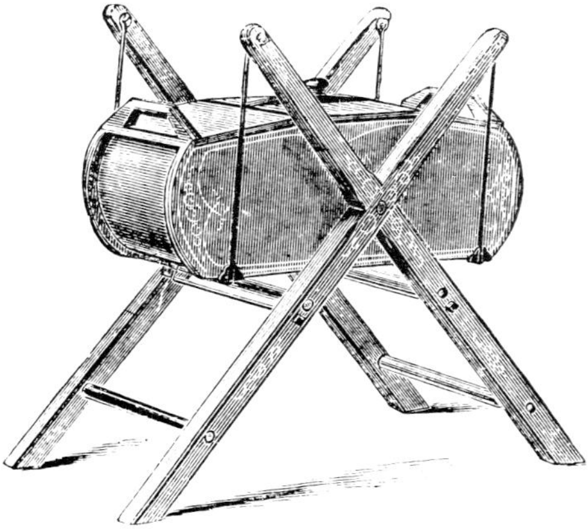

Davis oscillating churn, c. 1884
made in Vermont

Small metal churn

BUTTER CHURNS, the premier of dairy collectibles. It seems that man's imagination went wild when it came to inventing the churn, for they take so many different forms. Shown here are but a few.

Most churns were designed to churn only about half full.

Rectangular churn patented in 1868 was made by Cornish, Curtis & Green, Fort Atkinson, Wis. Came in seven sizes.

Small wood stave dash (dasher, stomper or plunger) churn is only 14 inches high.

White cedar churns came in several sizes and have been made for many years with little change in design.

Wood stave dash churn with metal bands. Many early stave churns made by coopers had split reed bands.

Union brand, three-legged diaphragm churn c. 1880's

A rare all-tin dash churn

Barrel-shaped dash churn

Barrel churns made to tumble on stand. Came in 10 sizes — 5-120 gallons.

This churn, with a thermometer set in its side, was made about 1857.

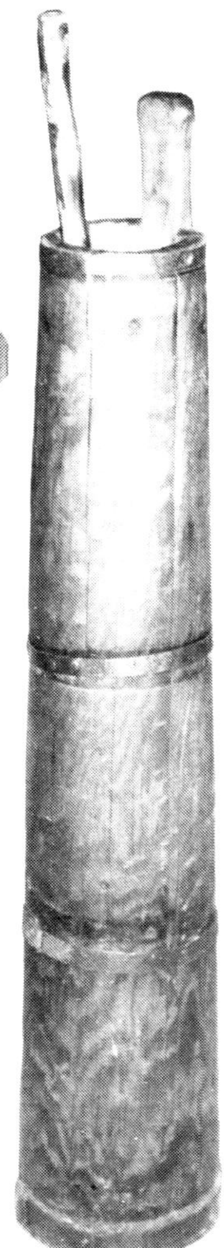

Very primitive churn has one stave longer at the top for a handle.

Sturges all-steel churn. Tinned steel barrel painted blue, stand in red. c. 1916

Blanchard diaphragm churn came in 5 sizes. Very popular in the 1880's.

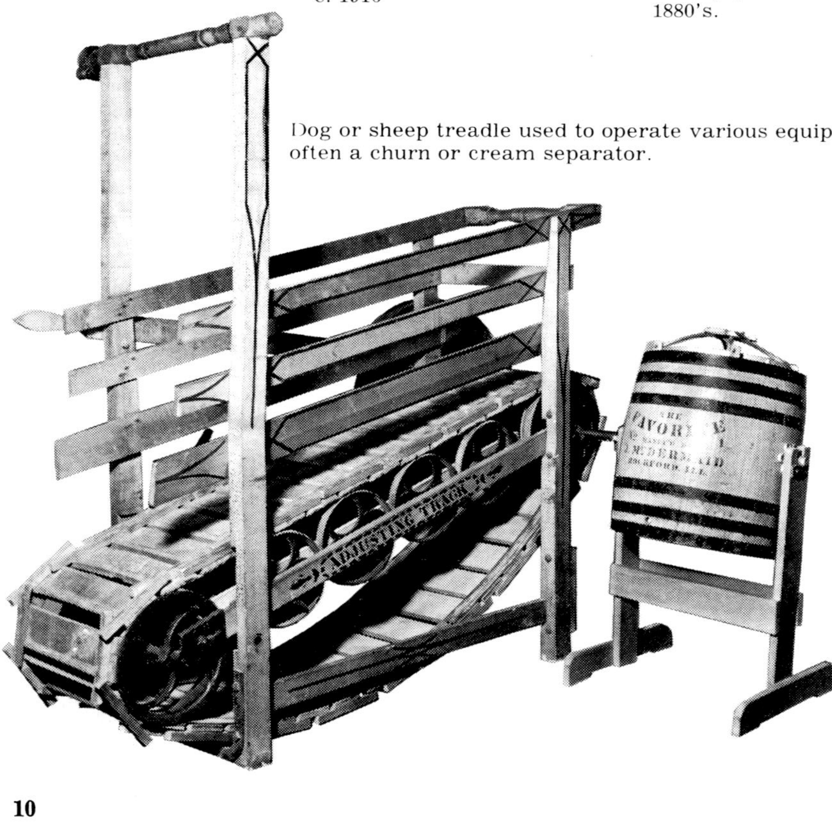

Dog or sheep treadle used to operate various equipment, often a churn or cream separator.

Dazey glass jar churn

Gem Dandy churn

Glass jar churn
with bail.

Newer model Dazey churn

Churn top
made to fit
on common
canning jar

STONEWARE CHURNS shown on these two pages are but a sampling of the churns available.

Many are found without tops or have new wood replacements. Tops broke easily, and some tops were sometimes cast aside as a nuisance or when the churn became a storage box.

5-gal. churn from Ohio region

4-gal. brown speckled churn

3-gal. churn, Wis. or N.W. Illinois

2-gal. Red Wing churn

4-gal. brown churn, southern Ill. region

Stoneware jugs were used for a variety of dairy-related products. This one made in Roseville, Ohio.

3-gal. churn with unusual design

4-gal. churn Lazy eight design, Ohio-Indiana region

5-gal. eagle design, Tennessee, Ill.

Simple 2-gal. stoneware churn, possibly Milwaukee, Wis.

CHURNS on this page are most unusual and very rare.

This churn hangs on a wall and swings to and fro by working handles. Was made by Labor-Saving Pendulum Churn Co., Bangor, Maine.

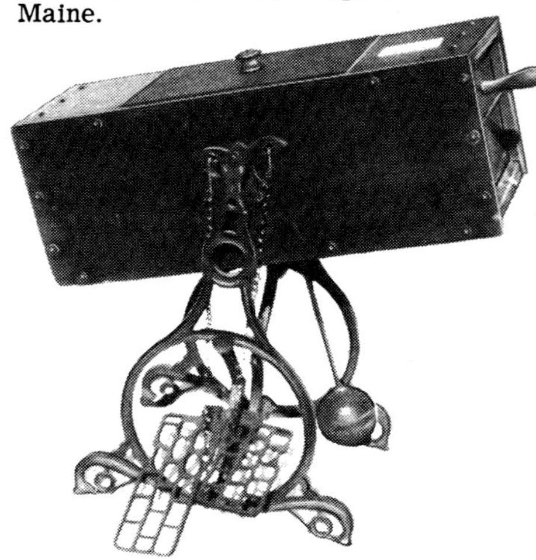

Foot-peddled churn sets in a metal frame. Note the counterbalance ball below that assists the swinging.

Bowl-type church fits inside a washing machine, and its agitator churns the cream.

J.P. Kelso's Railroad Churn, patented April 13, 1866. Made by Garten and Milligan, Bloomfield, Ind. It rocks back and forth on small train-like wheels.

Butter worker patented in 1875 by A. H. Reid of Pennsylvania came in 7 sizes.

BUTTER WORKERS were used to work the butter after churning to remove the whey and to work in salt.

Lever worker to the right was first made in 1877 by Cornish, Curtis and Green in 4 sizes.

Farm women commonly used simple wood bowl and ladle.

1895 advertisement for butter color.

BUTTER MOLDS AND PRINTS make beautiful collectibles. Many have one-of-a-kind, hand-carved designs. At one time, it was possible to order molds with handsome prints of any design. Butter molds generally came in quarter-pound, half-pound, one-pound or two-pound sizes.

Shown from upper left to right: sheaf of wheat mold and print, stylized plant print, cow print, small abstraction mold and print, strawberry (or heart) print, and small sheaf of wheat mold and print. Butter prints generally fall into three categories: factory, craft shop or homemade.

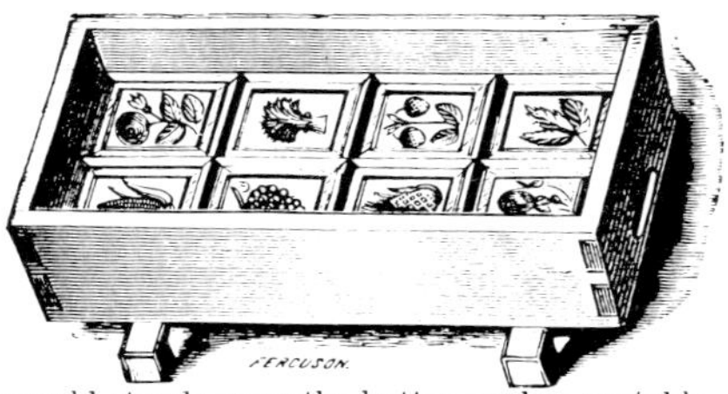

COMBINATION BUTTER PRINTS.

The cut here shown represents No. 6 print in its mould, which holds 2 lbs., and is divided into 8 cakes, ¼ lb. each, with a cutter which goes with each print. The butter is pressed into the mould on to the face of the print with a paddle, which is also used as a strick to strike off surplus butter. In filling, the mould stands upon the butter worker or table, and the butter pressed into the mould with the left hand on top of paddle. Strike from center of mould.

Old advertisement for combination butter prints.

Butter mold and print
with a swan design

Sheaf of wheat

Two acorns and leaf

Abstract design

Pineapple in the common half-pound size and uncommon one-pound size.

Eagle print

Eagle print

Print with stylized plant

Primitive fruit or flower?

Pinecone or natural plant design. On the right side view that shows handle and the normal one-piece construction of prints.

Mold and print with letter H makes 8 pats at once. Just cut apart.

Stylized plant print

A half-pound print of butter set in an old butter dish adds a nice touch to a special occasion. Butter must be soft enough to mold. Chilling the mold in ice water before and after pressing in butter seems to help release butter and retain a good print.

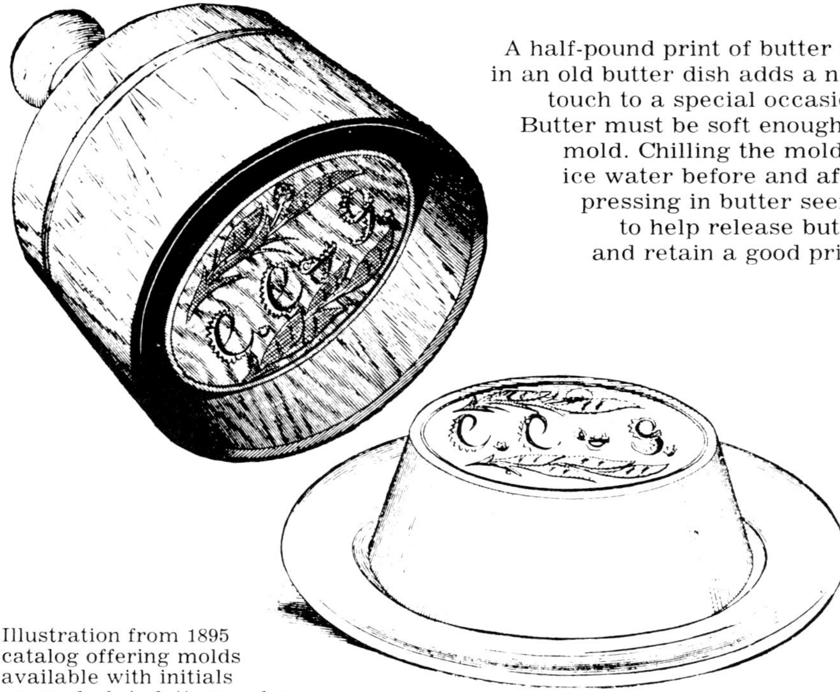

Illustration from 1895 catalog offering molds available with initials or word of six letters or less.

W. C. Freeman's four-part mold was patented in 1898. Has rotatable plate to simplify filling and ejection of butter.

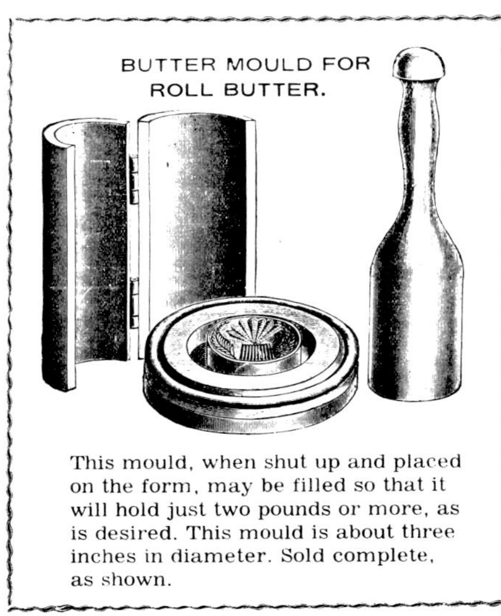

1895 advertisement

BUTTER LADLES came in a variety of shapes and sizes. Below is a butter tray with ladle used to handle butter from the churn before packing, sometimes used instead of a worker.

BUTTER AND CHEESE DISHES

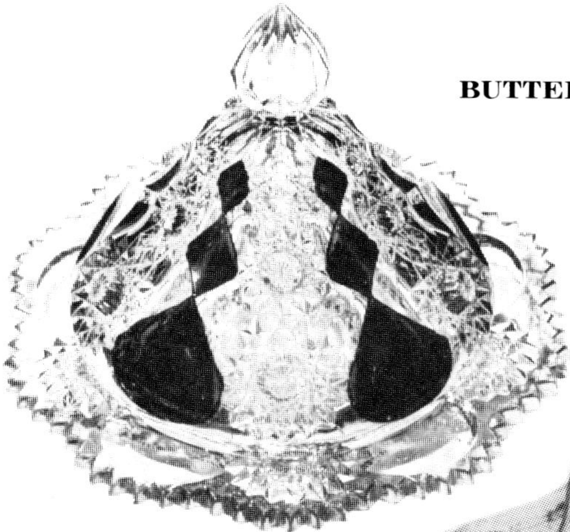

At left is an old, pressed glass butter dish. The dark areas are a deep red color and the rest of the dish is clear crystal.

Cheese dish at the right has an unusual cover shaped like a wedge of cheese.

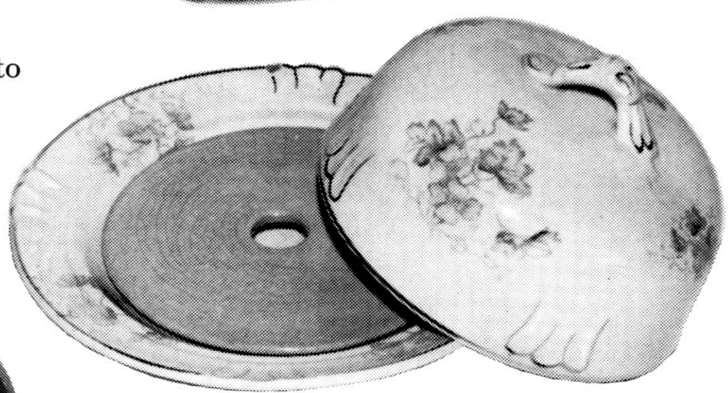

Most cheese dishes have a bottom insert to set the cheese on as shown here.

Butter dishes can be used in combination with a wooden mold to produce an attractive display of butter.

Many of these dishes were made during the Depression era and are highly collectible today. Most of them held about one-half pound of butter. At top left, center and bottom left, patterns include Pink Sharon, Pink Windsor and Green Cubist. Cheese dish second down from top left was given as a premium by Kraft. From top to bottom on the right are Green Princess, Pink Cherry Blossom and an early design in pressed glass.

BUTTER PATS
These little dishes were used to serve an individual butter "pat" at the table and were made of many different materials including china, ironstone and silver. Some were inexpensive and others were included in sets of fine china. Restaurants as well as homemakers used them.

Old eight-sided gift box for assorted cheeses.

Crock marked Red Wing on bottom

Western Stoneware with blue band

Plain butter crock

Cheese jars have been made for many years. Two shown made within last 15 years. Very collectible with printing on them.

STONEWARE CROCKS

"Old Sleepy Eye" crock made by the Weir Co., Monmouth, Ill., in 1903

Western Stoneware Co. crock came with a top. Made in the late 1920's.

Below are 4 crocks made in Red Wing, Minn., two with Hazel brand, small Land-O-Lakes and large 20-pound Newman & Son crock.

MILK AND CREAM CANS

Cream cans in 6- to 12-quart sizes make beautiful dry flower vases. Cans were generally used to handle cream which is only a small portion of milk. A dairyman with a separator and a small herd needed only a few of these small cans to handle his cream.

The small milk or cream cans shown below were used to handle the milk and cream for the farm household. In the days of street vendors, housewives would get their milk in these. Similar cans are used on dairy farms today for the house milk.

Green and white enameled can on left was made in Austria. Middle can is gray enameled. Tinned can on right is a larger 4-quart size.

A variety of milk and cream cans used on dairy farms. Largest one shown is 12-quart size. Many cans had covers with inverted bottoms which could be forced down on cream to drive out air and prevent churning when in transit. Chains kept covers from getting lost.

MILK CANS

This advertisement for Sturges products in 1900 gives an idea of the variety of cans available. There were a number of manufacturers. The 10-gallon can became the most popular, although in some areas you can find an abundance of 8-gallon cans. Milk cans are still in use on many farms today.

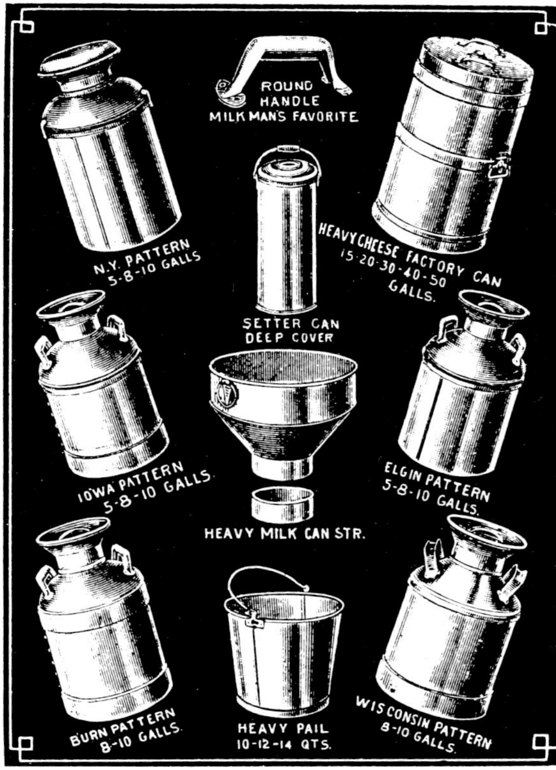

Can at left is an older one. Notice the rivets used on seams and to attach handles. Newer cans are generally welded and seamless. The can above is an unusual railroad can. With its tapered sides, only the bottom band touched the adjacent cans; this kept them from getting battered.

Above is a cream setting can made in 1879. Center column was for cold water to help cool milk. Old can at right shows a different handle. In 1912, Creamery Package Co. alone offered eight different handles.

FACTORY AND JACKETED MILK CANS

These cans were in wide use when herds were small and only one or two cans could handle all the milk from the farm. Hoists were needed at both farm and creamery to handle these large cans. They came in sizes of 15 to 60 gallons; 20- or 30-gallon sizes were most popular. Some cans had wood or metal jackets to help insulate them and keep milk or cream cool. Most came with plunger floats to keep milk from churning.

From 1895 catalog.

Large jacketed can made by J.G. Cherry Co.

Old advertisement showing factory cans being delivered to plant.

This 20-gallon wood jacketed can is in like-new condition. Creamery Package took over Cornish, Curtis and Green products in 1898.

Drawings of jacketed cans from the 1880's.

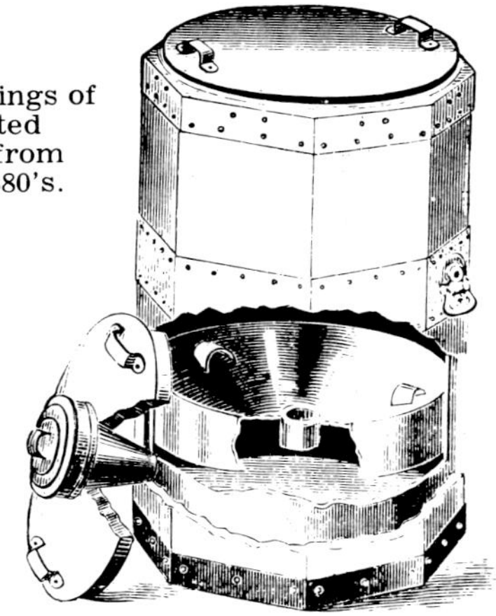

This can has cast in the bottom "Allen Bros. Fort Atkinson, Wis. Design Feb. 20, 1887."

DEEP SETTING CANS were used to cool milk and let the cream rise. They sometimes were immersed in ice water. Some had windows in side to measure amount of cream.

COW HOBBLES are used to keep a cow from kicking during milking. They hooked over hocks; chain was drawn up tight.

These cans or pails shown to the right are used in a variety of ways, such as transferring milk from buckets to large cans or to milk room coolers and tanks.

MILK PAILS AND BUCKETS
The half-covered pail
was designed to help keep foreign
material out of milk. The primitive,
wood-staved bucket is made of cedar.

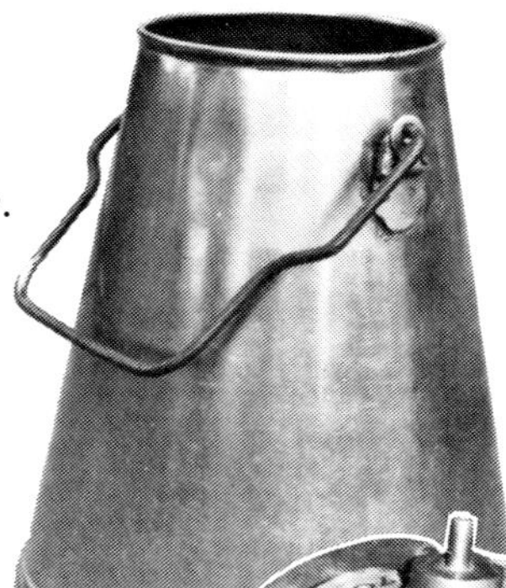

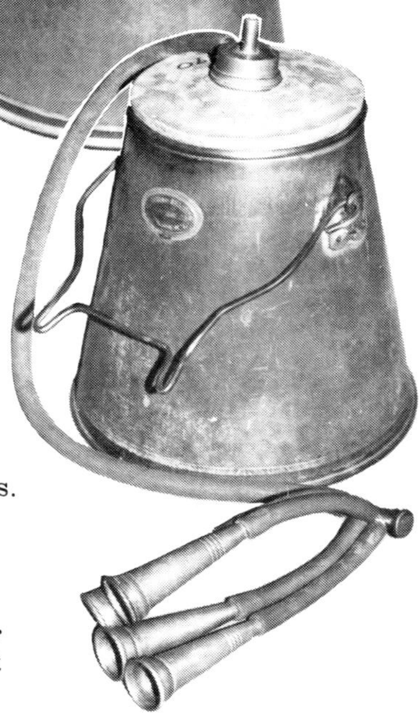

These are milking machine buckets. The one above is made out of aluminum. At upper right is a stainless steel bucket. To the right is a 1910 Hinman milker. These buckets are ideal for hauling water for yard and home. They don't rust or spill easily.

1940 combination milking unit

Milking machines made like this with compressor, motor (choice of electric or gasoline), claw and cups all on one cart, including the milk can, were being sold in the late 1930's. World War II stopped their production.

Mehring foot-powered milker was invented in 1891 and was still being offered for sale in 1925!

Picture from 1925 ad shows man milking two cows with this milker.

FORDS MILKER (c 1927) Operated by either electric or gas engine and ran on a track behind the cows. A copy of an old magazine ad helps support a display.

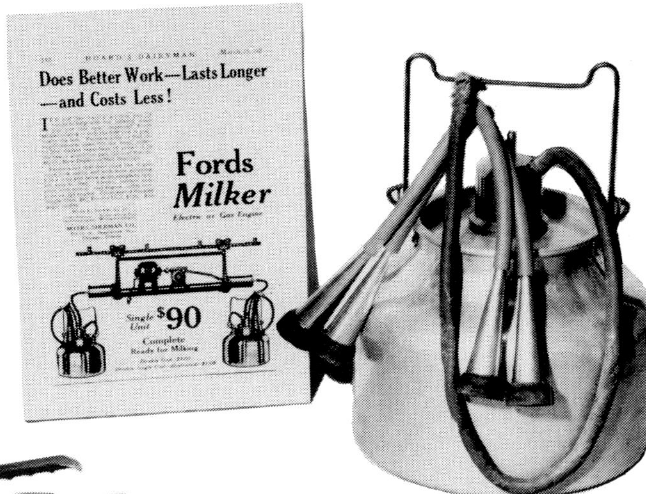

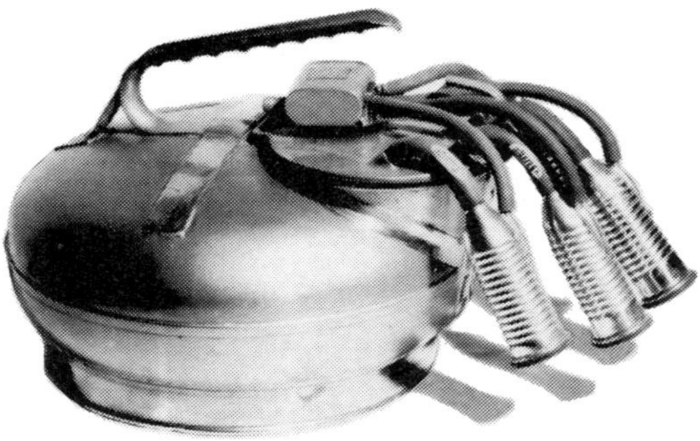

PINE TREE SURGE MILKER (c 1923) Hung from strap over cow's back. Often called 'tug-n-pull' style.

MILK COOLER-AERATORS. Milk was poured down the conical sides. Cold water is run through inside. Aeration removes animal and barn flavors.
The two above were invented by Hans K. Christensen, Fort Atkinson, Wis.

STRAINERS came in many shapes and sizes. They make nice lamp reflectors and hanging plant containers.

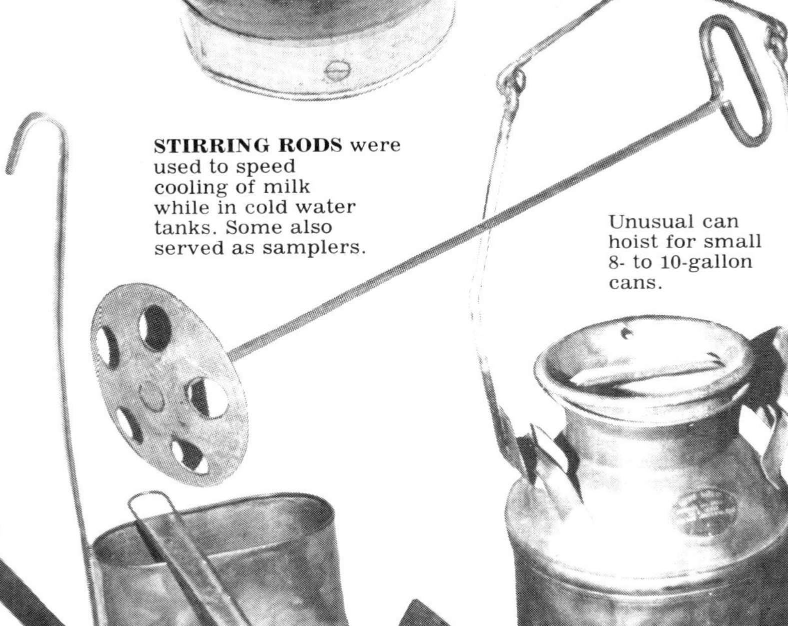

STIRRING RODS were used to speed cooling of milk while in cold water tanks. Some also served as samplers.

Unusual can hoist for small 8- to 10-gallon cans.

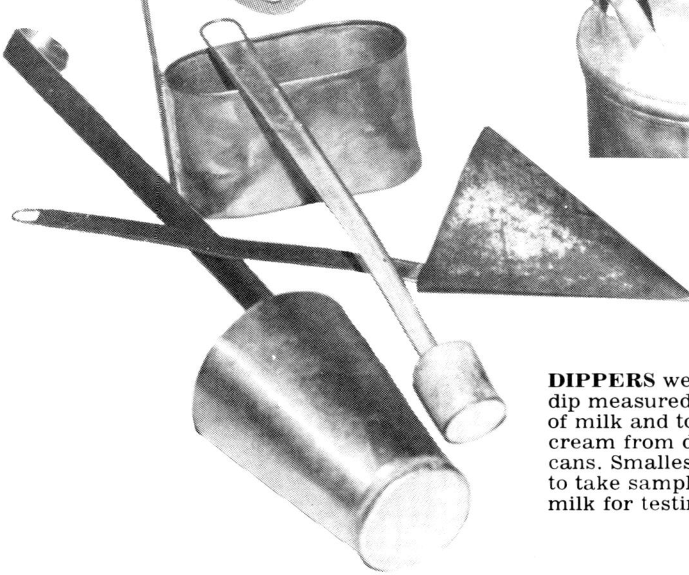

DIPPERS were used to dip measured amounts of milk and to skim cream from deep setting cans. Smallest one was to take sample of milk for testing.

MILKING STOOLS

At some time in history (no one seems to know when), man decided to sit on a stool to do the milking. Since then, stools of all shapes have been designed. Here are but a few that have been saved.

Simple, wooden, four-legged milk stool.

Creamery Package manufactured stool.

Three-legged stool made from fork of tree.

One-legged wooden stool.

Unusual wood stool with hoop to hold bucket.

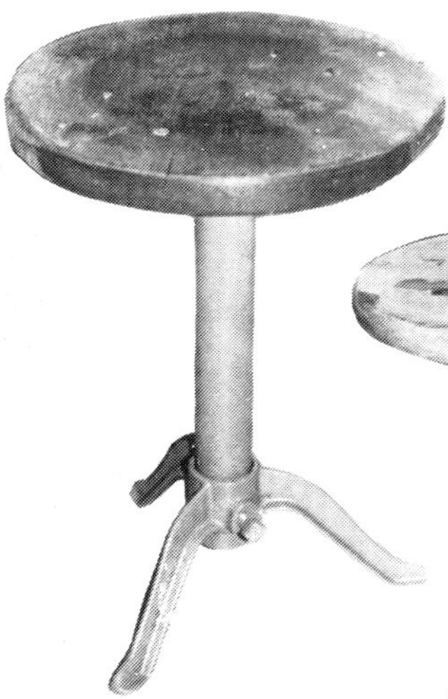

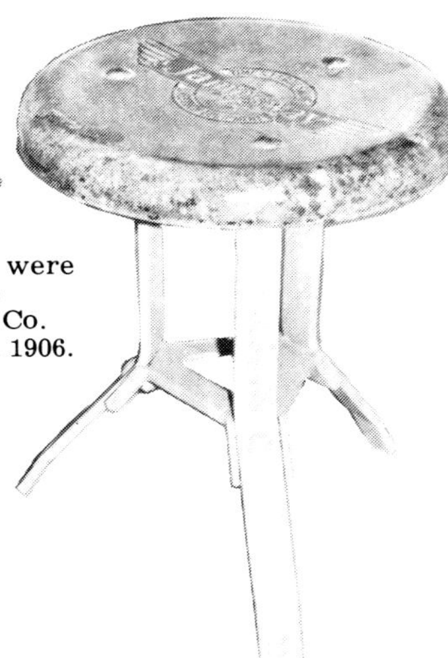

These stools were made by the James Mfg. Co. beginning in 1906.

MILK BOTTLES
Bottles have always held the interest of collectors. Shown to the left is an early Thatcher bottle. These were first produced in 1884. Amber bottle with tin top is at upper left. At center is a cream separator bottle from 1930 — the dent helps hold back cream while pouring off skim. Bottle with tin handle was made by A. G. Smalley, 1898. In foreground is a half-pint Borden Eagle Brand bottle. At lower right is a quarter-pint bottle.

FIRST GLASS MILK BOTTLE This advertisement for Warren Milk Bottles appeared in the U.S. Dairyman in December, 1883. Many of the bottles shown below have applied color labels which first came in use in early 1930's. In center and lower right are two Baby Face cream top bottles. At upper left is a cream top bottle with cream spoon. This would pinch off milk while cream from neck was poured off. Note 1925 bottle at left with copper "Verifine" token implanted in it.

MILK BOTTLE CAPS from old salesman sample book showing the variety of caps available. Caps like these have a wide range of colors, suitable for framed groupings.

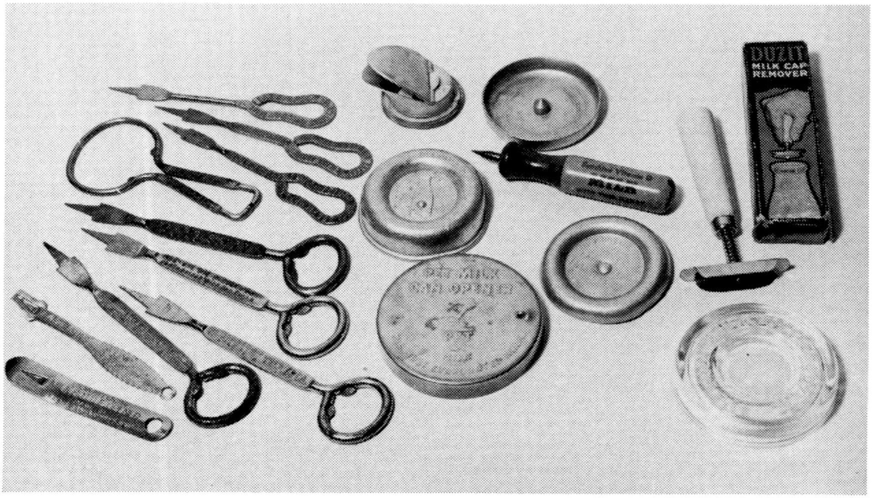

BOTTLE CAP OPENERS can make a unique collection by themselves as shown by this group. The sharp cap picks shown to the left punched into the cardboard caps to remove them. Large, round opener in center is for condensed milk can. Small round metal openers would serve as caps after removing cardboard caps. At lower right is a glass cap used after the bottle is opened.

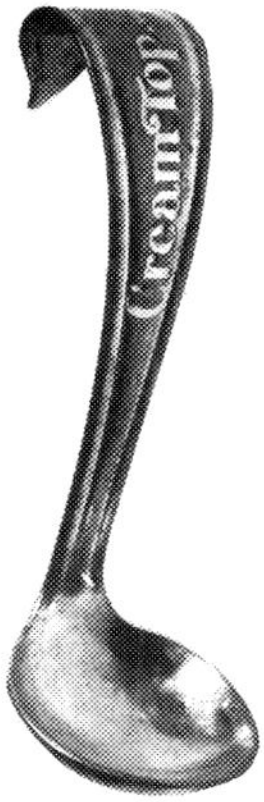

DELIVERY BASKETS for milk bottles came in a variety of shapes

CREAM TOP SPOONS were used in cream top bottles to block in milk when pouring off cream.

CREAM REMOVERS were means of siphoning off the cream without disturbing cream line.

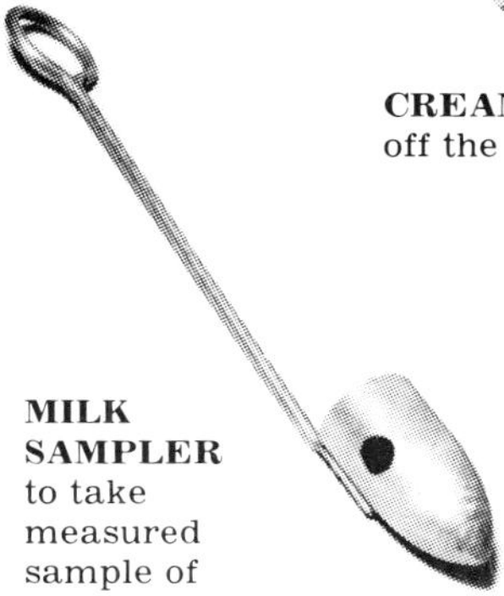

MILK SAMPLER to take measured sample of milk

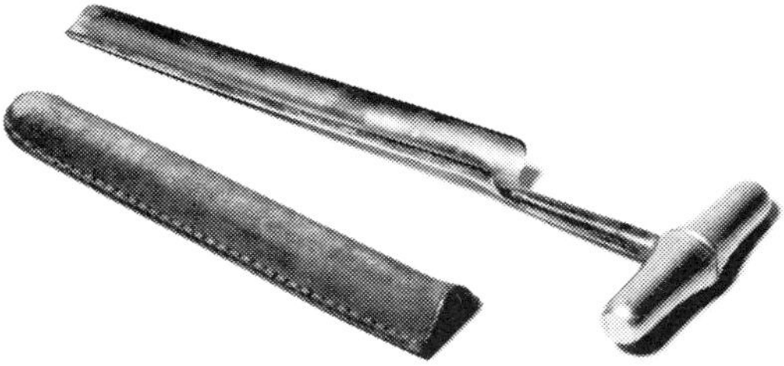

CHEESE TRYER shown with leather sheath is used to take core sample of cheese.

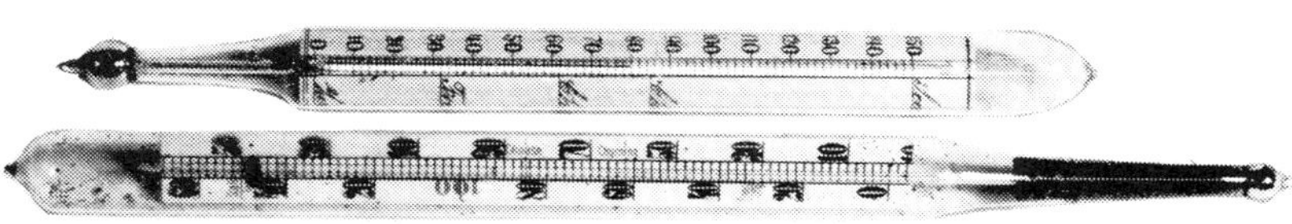

THERMOMETERS are 7 to 10 inches long. They are marked out at 150° for scalding, 85° cheese, 62° churning and 32° freezing.

MILK AND CREAM PITCHERS. These have been made in almost everything imaginable — glass, china, stoneware, silver plate, pewter, wood, etc. Every size, color and shape can be found. Many resemble animals and birds, even people. Shown below at upper left: Red Wing gray line pitcher; upper right: sponge or spatterware pitcher — midwestern; next down on right: Chelsea pattern, England c. 1880's. The rest are cream pitchers from dish sets.

On the left is a primitive milk pitcher made of wood. On the right is an old stoneware pitcher that is about 6 inches high, probably from the Ohio region.

Beautiful Moss Rose ironstone pitcher. The one with Jersey cow on it was given as an award.

SMALL CREAMERS are collected by many people. They can be displayed charmingly in a small space. The simple ironstone creamers used by hotels and restaurants are the most common.

BABCOCK TESTERS. A simple and accurate method of testing butterfat content was invented by Dr. Stephen Babcock. Testers were first marketed in 1890.

Sulphuric acid is added to milk in a graduated bottle; then mixture is whirled until butterfat comes to top of test bottle where it is measured.

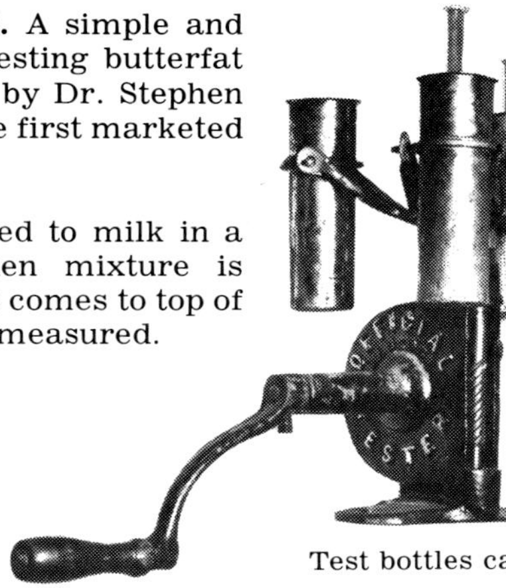

Official four-tube tester sold by Creamery Package 1912-1924

Two-tube tester. Bottle tubes on all these testers are brass. c. 1910

Test bottles can be purchased today.

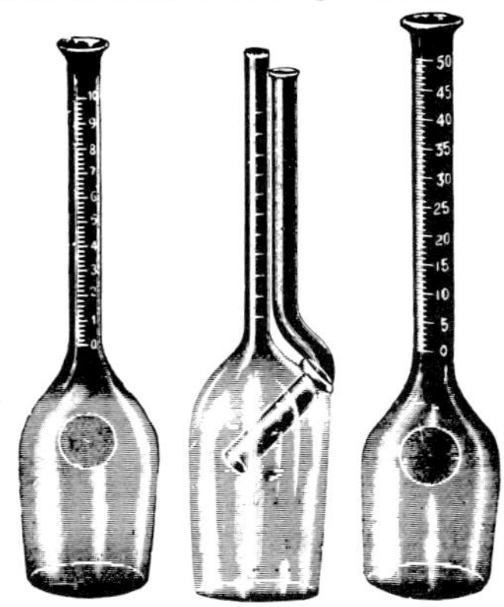

Milk Skim milk Cream

Standard Hand Tester complete with shipping box and instructions was sold by Sears.

Wood frame Babcock tester shown in an 1890 catalog — the year testers were first marketed.

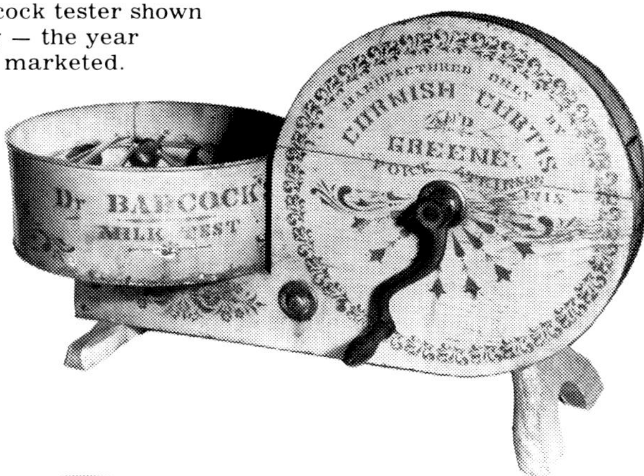

Cornish, Curtis & Green tester. c. 1895

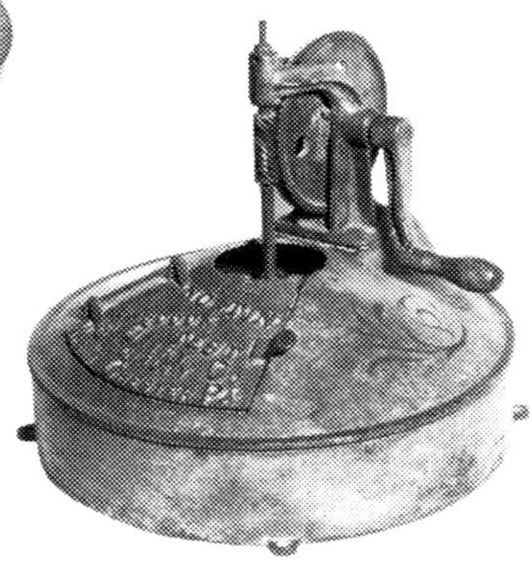

Red-colored "No-Tin Tester" made by Dairy Specialty Co., West Chester, Pa. c. 1915

20th Century tester made by Creamery Package Co., Chicago, Ill. c. 1912

Facile tester pat. April 10, 1900

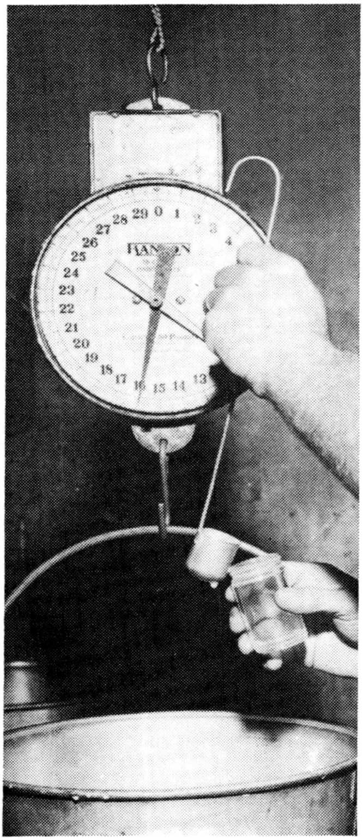

A scale used in conjunction with a sample for fat test is still in use on farms today.

MILK SCALES

For many years weighing milk has been a check on a cow's production. The steelyard was probably the first scale used.

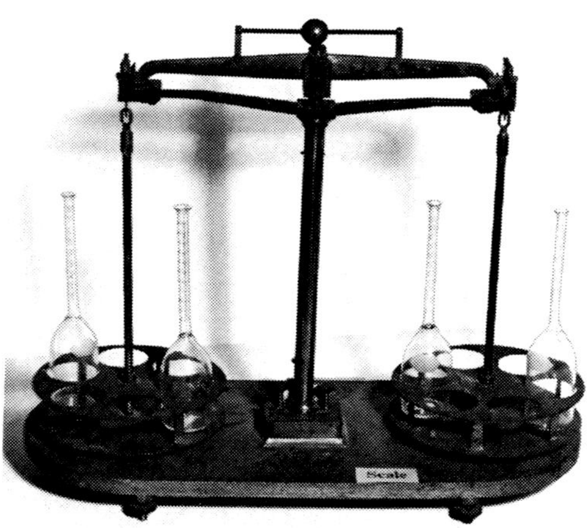

CREAM TEST SCALES used in plant labs. One shown holds 12 Babcock test bottles.

Brass milk scale 10 inches high

CREAM SEPARATORS were partly responsible for the dairy industry's rapid growth in the late 1800's. They offered a quick means of separating cream from milk.

1907 Simplex separator made by the Burrel Co.

Tabletop Montgomery Ward separator

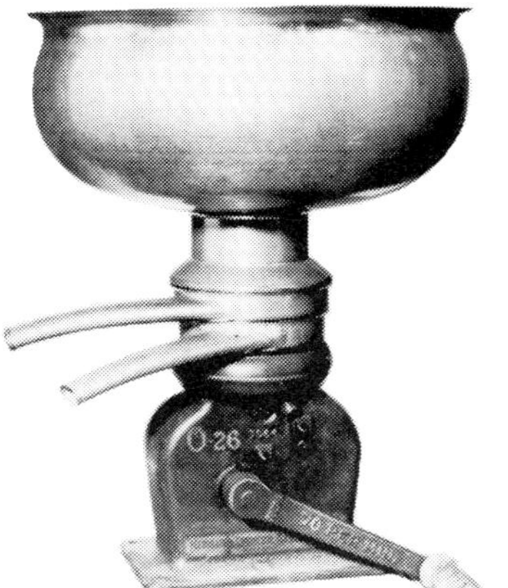

DOMO JR. TABLETOP SEPARATOR
Most tabletop models are found in excellent condition, because they were stored in the home.

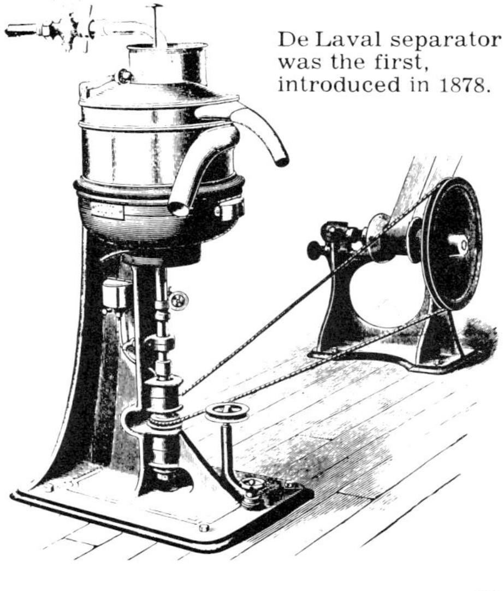

De Laval separator was the first, introduced in 1878.

DELAVAL ALPHA NO. 1 BABY SEPARATOR
is an early one and quite rare.

DeLaval was the largest and most prolific of the separator companies; many different models were offered over the years.

DELAVAL DEALER
PARTS CABINET
would be the highlight of a cream separator collection.

GRAVITY SEPARATOR has two windows to follow cream line as milk is withdrawn from bottom.

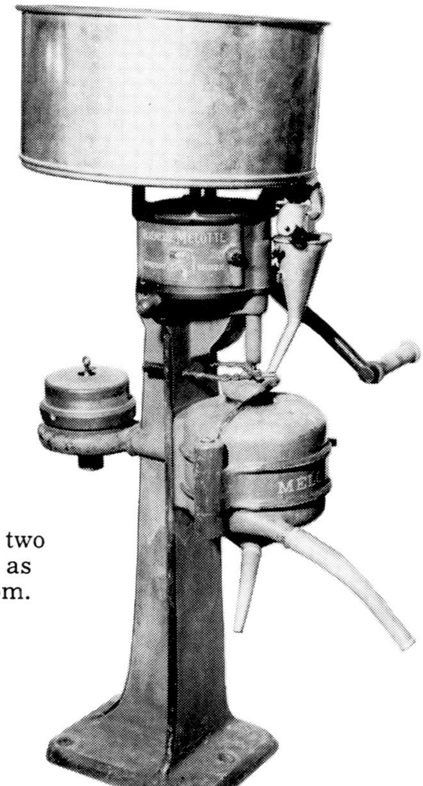

MELLOTE SEPARATOR Imported by Babson Bros. in 1913 to mark their entry into the dairy equipment business. This model has a nice red color trimmed in black with gold lettering.

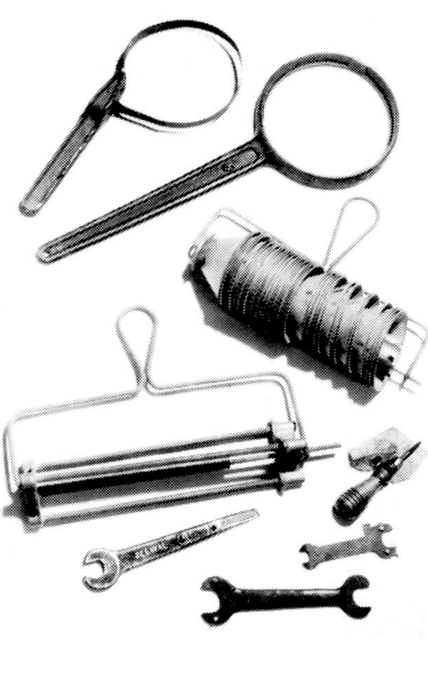

CREAM SEPARATOR TOOLS Separator collections would not be complete without all the tools and accessory parts.

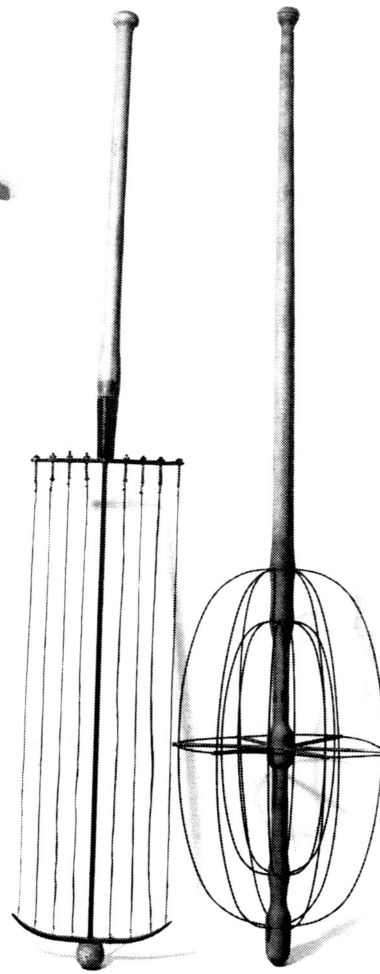

CHEESE PRESSES are rare finds and the copper Swiss cheese kettles even harder to find. Author had to use old catalog drawing. Listed were sizes in 100 to 450 gallons. The 6-foot curd knives (right) are sometimes called harps and used in kettles.

CHEESE BOXES are good display bases and storage boxes.

CREAMERY AND CHEESE FACTORY ITEMS. Shown is a tinned strainer, brass strainer with ornate iron handle, large dipper, curd rake, small and large butter packer, longhorn cheese hoop, butter boxes, and butter spade. Rake, spade and strainers would make nice wall hangings along with the harps on opposite page.

PROCESS CHEESE BOXES have many uses in the home. Spice boxes and dry flower containers, recipe files and miscellaneous kitchen tool boxes are but a few uses for them. They came in sizes from 2 to 5 pounds.

DAIRY PANS
Milk was placed in these simple, tinned pans to let the cream rise. Cream was then skimmed off. There were three ways to separate cream from milk: dairy pans, deep cans and ice-centrifugal separators.

Pans also were made of glazed earthenware and some milk pans may have been made of glass.

Many old drawings show a large number of dairy pans in them.

BUTTER BOXES make nice collectibles. Box at upper right held five 1-pound packages. Butter coloring sometimes came in small wooden boxes also.

ICE CREAM SCOOPS are very collectible, some with their intricate mechanisms and others in contrast with their simplicity. Some of these date back to the 19th century.

Colorful, framed glass sign is but a sample of signs to be found.

Mixer and bowl for whipping cream was used in promotion for a creamery.

PEWTER ICE CREAM MOLDS were popular in the early part of this century. A 1912 Creamery Package catalog listed 500 individual metal ice cream molds all the way from the ace of spades to a yacht. Also listed were over 80 larger molds used by hotels and restaurants. Shown here from left to right starting at top: Half-peeled orange, turkey, pumpkin, Santa Claus by chimney, Santa Claus carrying gifts, rabbit, heart with cupid, rose, a different turkey, cow and Washington on a hatchet.

ICE CREAM FREEZERS came in 2-, 3-, 4-, 6-, 8-, 10-, 12- and 14-quart sizes. Larger ones were made for commercial plants. The one on the right is a Blizzard freezer from the 1890's. Below is a smaller White Mountain freezer and on the right is a metal Snow Peak freezer.

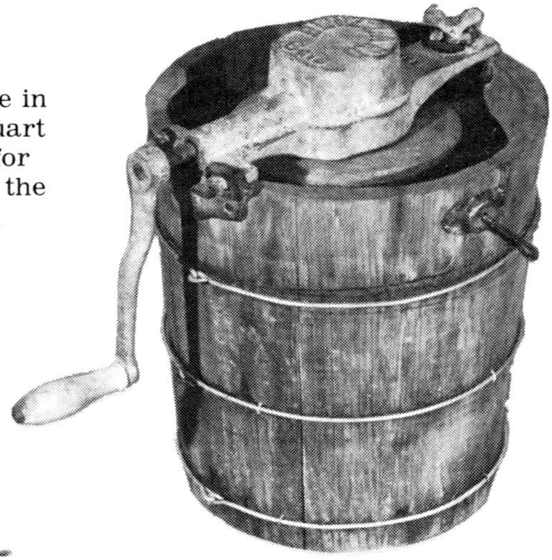

Shown at left is an unusual insulated ice cream container made of light cardboard. On the right is an ice cream can which was generally kept in an insulated canvas carrier.

A very simple tinned freezer from 1920's

COW FIGURES

The cow has been immortalized in many forms, such as statues, paintings, drawings and ceramics as shown below. The cow figure has made a good subject for children's toys, also.

Bisque figurine is over 100 years old.

Except for stuffed figure to the rear and small glass cow in lower right, these are all ceramics.

COW PITCHERS, whether new or old, have an attraction for the collector. Many were designed more for decoration than for use.

Painted tin toy cow. Pump its tail and the eyes open and shut while it moos.

Cow-shaped spoon rest

Handcrafted Christmas cow ornaments. The two lower ones were made from bread dough.

COW WEATHER VANES were furnished for many years by barn equipment companies such as Jamesway, Hudson, Starline, Louden, etc. They were installed on top of the barn ventilators (metal cupolas) or lightning rods. Most figures were made of zinc.

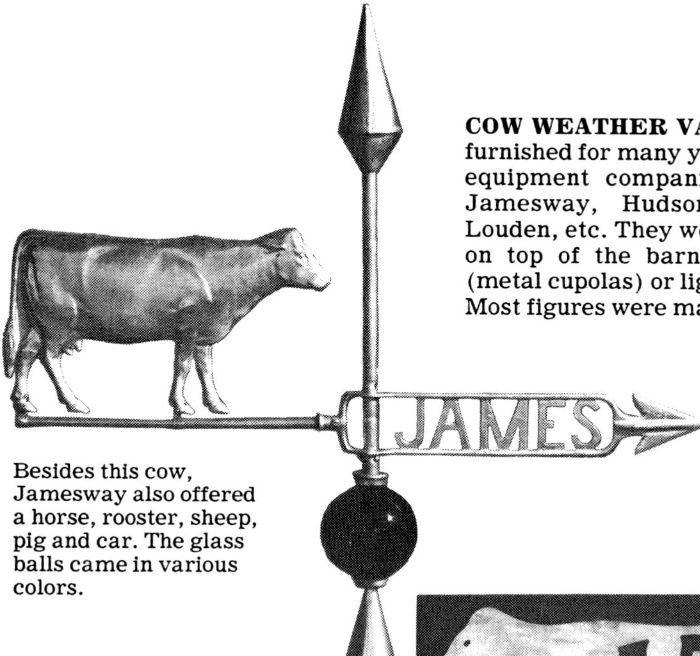

Besides this cow, Jamesway also offered a horse, rooster, sheep, pig and car. The glass balls came in various colors.

Author does not know what AAA represents.

This cow model with horns is seldom seen.

This cow is seen mostly on lightning rods.

TIN PRODUCT PREMIUMS such as these painted tin cows and calves gave many an hour of pleasure to a farm boy or girl early in this century.

1908 separator match stick holder shown at upper left was for Mom. New matches were kept in separator bowl; burnt ones went in bucket. The holder was hung by the kitchen wood stove. A 1906 tray is at upper right. The Holstein calf and cow are from 1909.

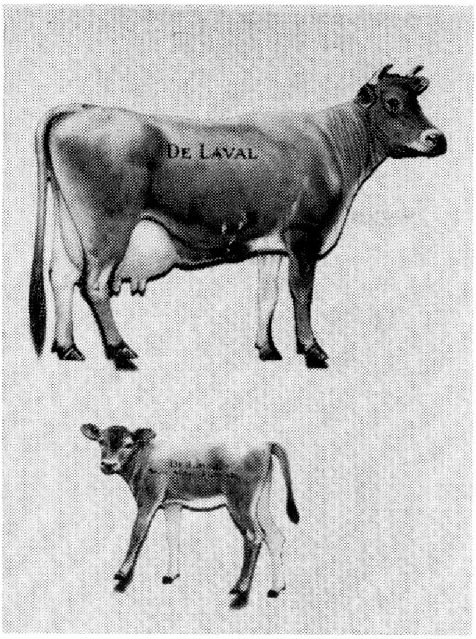

Above is a Jersey cow and calf from 1922. At top right is a 1919 Guernsey cow and her DeLaval skim milk baby with the original envelope they came in.

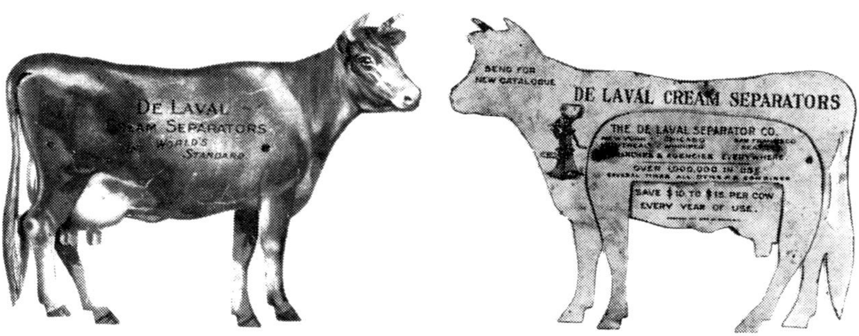

This unusual Jersey cow from 1909 was made in two parts. Shown here are both sides. The DeLaval cows are about 3 inches high at top of back.

There are many other tin product premiums to be found such as this "Success" manure spreader printed tin tray.

COWBELLS came in a variety of shapes and sizes. Where dairy cows were allowed to forage over wide areas of wilderness, bells were hung from their necks to aid in finding them for milking.

Swedish bronze cowbell with clapper made like a small bell.

Swiss cowbells, the "granddaddies" of all cowbells, are worn by Brown Swiss cows (shown above) in the annual Wilhelm Tell Pageant at New Glarus, Wisconsin. There was an older, tulip-shaped bell, also.

Commercially made bell with some of the original label still visible.

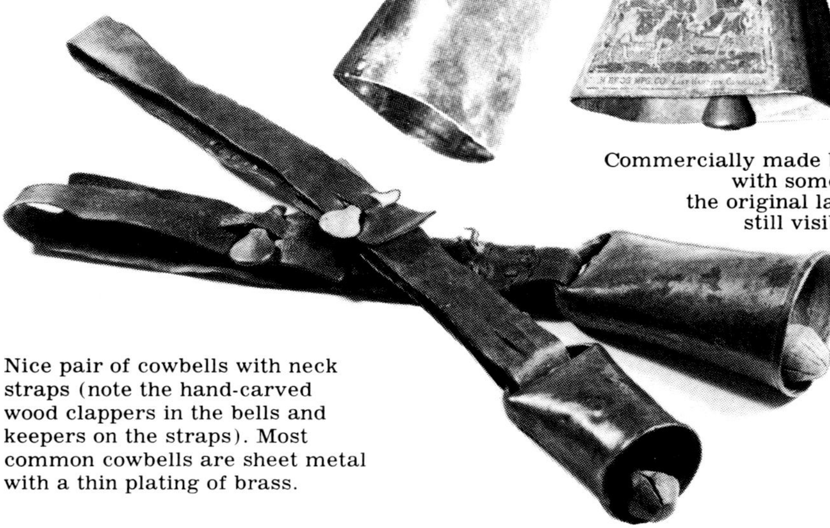

Nice pair of cowbells with neck straps (note the hand-carved wood clappers in the bells and keepers on the straps). Most common cowbells are sheet metal with a thin plating of brass.

ELSIE COLLECTIBLES: Elsie doll, Beulah doll, Elsie game, ice cream box, Beauregard mug, Elsie creamer, Elsie glass, Elmer sugar bowl and Elsie creamer set, large Elsie sign, comic book and blotter. There are many more Elsie and her family items that were produced by Borden.

CONTAINERS as accessories can add an inexpensive touch of color or interest in setting the mood of a country-style room.

Many were made of cardboard, metal, plastic or combinations of these.

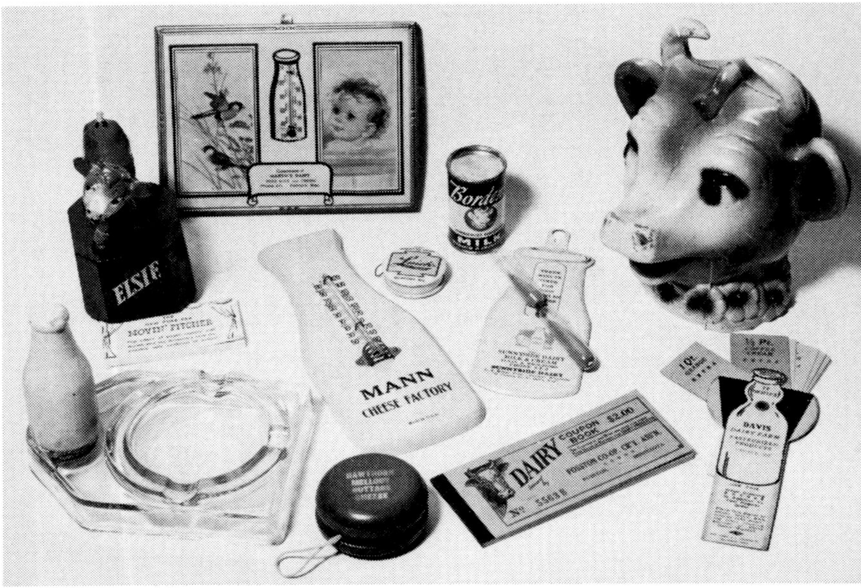

PROMOTION PIECES given out to advertise dairy products. These include such things as ashtray, toys, thermometers, tape measure, egg timer and "Elsie" bank. Routeman's coupons and product listings are shown at bottom right.

BUTTONS used to promote milk and dairy related products.

HITCH HIKER CREAMER
hooks on saucer but apparently didn't catch on with the public.

SALT and PEPPER SHAKERS
are very popular and can be found with dairy themes.

PLAYING CARDS
are collectibles that can also have dairy advertising.

ALMANAC MATERIAL
can be found in calendars and mfg. yearbooks.

OLD POSTCARDS of milk delivery wagons scenes at creameries and cheese factories, and dairy farm views are fun to look for.

OLD ADVERTISEMENTS AND PICTURES. There are many old and beautiful illustrations that can be used to complement dairy collections.

DAIRY RECIPE BOOKS seem to keep showing up at flea markets and auctions. Some of the old recipes are still the best.

BUMPER STICKERS are a modern collectible with their varying slogans to promote dairy products. **COW GREETING CARDS** are not only fun to collect but also can be a good way to share your interest in the cow.

FARM SIGNS help support a collection and do not take up floor space.

CARDBOARD CONTAINERS AND DISPLAY ITEMS. Ice cream, cheese, yogurt and milk containers made of waxed cardboard are colorful and inexpensive to collect. The cherry-topped sundae cut-outs are for soda fountain display. The Yogurt Amiot jar is ceramic.

BUYERS' PRICE GUIDE TO DAIRY COLLECTIBLES

Prices listed reflect the value of items in **good-to-excellent** condition. Chips, cracks, rust and missing or broken parts can drastically lower the price. Most artifacts, if truly old, will show some wear, and this only adds to their beauty. Price guides should only be that—**a guide** !

Often the determining factor is how badly you want an item or how willing the seller is to sell it.

Locality has a great deal to do with prices, also. Local brand names generally are more in demand. These prices are based primarily on sales reported in the midwest.

Page 7. Oscillating churn $350-500, square metal $65-100, rectangular $300-400, 14-inch wood stave (no top, primitive) $175-225.

Page 8. 24-inch wood stave churn with top $210-320, white cedar $100-175, all-tin dash $125-200, Union $325-400.

Page 9. Barrel-shaped churn $150-200, barrel on stand $125-190, thermometer churn $110-185, primitive stave $250-350.

Page 10. Steel Sturges churn $45-85, Blanchard diaphragm churn $190-275, treadle $725-800.

Page 11. Standard glass jar churns $45-60, brand name churns like Dazey $75-100, small fruit jar churn $35-45.

Page 12. Speckled churn $150-250, 5-gal. churn $200-300, 3-gal. churn $225-300, 2-gal. Red Wing $125-160, brown churn $90-155.

Page 13. Stoneware jug $30-50, 3-gal. wood top $125-150, 4-gal. lazy eight $190-375, 5-gal. eagle $200-375, small, simple churn $75-125.

Page 14. These four very rare churns $450-700? (Have not seen one sold.)

Page 15. Reid butter worker $175-225, Curtis butter worker $200-300, 15-inch wood bowl $12-20, ladles $12-25.

Page 16. Wheat mold and print $70-90, stylized plant print $60-75, small strawberry print $70-130, cow print $175-250, small individual pat molds $30-45, combination mold and print as shown in ad $75-110. (Many reproductions of all molds are showing up and difficult to spot.)

Page 17. Swan $75-85, wheat $70-90, acorns $60-80, abstraction $50-65, 1/2-pound pineapple $65-95, 1-pound pineapple $90-145.

Page 18. Eagle prints $140-200, stylized plant $60-75, fruit $55-70, pinecone $65-80, stylized plant $60-80, combination mold and print $75-110.

Page 19. Freeman's four-part mold $70-90.

Page 20. Butter ladles $10-20 depending on their intricacy (Watch for new reproductions!), butter tray $60-90.

Page 21. Pressed glass butter dish $60-75, cheese-shaped dish $40-85, round cheese dish $30-45, leaf mold and print $55-70, Cherry Blossom butter dish $60-70.

Page 22. Sharon, green $75, amber $45, pink $45, Windsor $40-45, Cubist $45-50, Kraft cheese dish $10-15, Princess, green $40-65, Cherry Blossom dishes $60-70, definite-pattern pressed glass butter dish $30-40, dishes of unknown pattern $10-20.

Page 23. Tea leaf (round or square) and moss rose pats $12-20, fine china pats $5-10, hand-painted pat $35-50, majolica (leaf shape) $40-70, pats from most china or ironstone sets $2-8, silver plated $15-20.

Page 24. Assorted cheese boxes $8-20, recent cheese jars $3-10, butter crock with blue band $15-30, plain crock $9-15 (marked Red Wing on bottom $20-35) Sleepy-Eye crock $200-250, blue printed crock (no top or handle as shown) $60-70, with top and handle $75-100, Hazel crock $125-150, Land-O-Lakes $35-55, 20-1b. crock $150-175.

Page 25. Green enameled can $45-65, gray enameled $35-55, 4-qt. can $25-30.

Page 26. Large can with cover $25-40, smaller cans $15-25, 6-qt. with cover $25-35, 12-qt. can $25-40.

Page 27. 10-gal. can $10-20, 5-gal. $18-25, 8-gal. $18-25.

Page 28. Riveted 8-gal. can $20-32, railroad can $40-60, cream setting can $40-60, 10-gal. DeLaval can $20-25.

Page 29. 30-gal. wood jacketed can $75-120.

Page 30. 20-gal. wood jacketed can $75-120, metal factory can $35-45.

Page 31. Cow hobbles $5-8, deep setting can with cover $15-20, no cover $5-10, deep pails $4-8.

Page 32. Covered pail $15-20, wood bucket $55-80, stainless steel $22-30, aluminum $10-15, Hinman milker $60-90.

Page 33. Combination milking unit $65-95 (less can). Mehring milker is a museum piece $475-550.

Page 34. Fords milker $80-125, Pine Tree Surge $45-70, cooler-aerators $40-60.

Page 35. Strainer $10-15, can hoist $12-18, stirring rod $4-8, brass dipper $20-25, conical shape $10-15, sample dipper $5-8, stainless steel $10-15.

Page 36. Primitive wood stool $15-25, three-legged stool $30-50, CP stool $9-15, one-legged stool $6-9.

Page 37. Stool with bucket hoop $20-30, wood-top stools $10-15, metal tops $5-10, strap-on stool (not shown) $40

Page 38. Milk bottles in middle row: amber tin top $30, bottle with dent $40, with tin handle $40, others $7-12, Thatcher $150. (Beware of repro's.)

Page 39. Embedded token bottle $18-25, cream tops $12-18, common bottles $6-9 (Locality has a great influence on price.), quart baby face $60-80, half-pint baby face $70-85 (Baby face bottles are being reproduced.) tin top $12-15, other bottles $7-12.

Page 40. Bottle caps $.25-.50, top lifters $7-10, lifter caps $7-12, glass or metal caps $8-15.

Page 41. Cream top spoon $12-18, delivery basket $18-25, cream removers with boxes $5-7, sampler $4-8, cheese tryer $8-12, thermometers $3-6.

Page 42. Red Wing gray line pitcher $150-175, spongeware pitcher $100-160, Chelsea $35-65, small pitchers from sets $4-25.

Page 43. Primitive wood pitcher $90-130, small brown pitcher $15-25, Moss Rose pitcher $100-125, Jersey cow pitcher $25-45, small creamers, plain $3-5, decorated $7-10, bottle with name $12.

Page 44. Four-tube tester $45-65, two-tube tester $35-45, milk and cream test bottles $4, skim milk test bottles $12, tester with box and bottles $55-80.

Page 45. Wood frame tester $450+, tester on stand $65-90, iron testers $40-65.

Page 46. Milk scales $20-35, brass milk scales $25-50, cream test scales $50-85.

Page 47. Ward's separator $70-110 (separating condition), Simplex $200-350 (attractive display piece), Domo Jr. $85-120. DeLaval separators vary in price from the common models to the rare $200-600 complete in good-to-excellent condition.

Page 48. Delaval No. 1 Baby $450-600, Delaval cabinet $450-650.

Page 49. Gravity Separator $35-65, Melotte $250-350, Tools $2-7.

Page 50. Cheese press $250-310, copper Swiss cheese kettle $550-700, cheese boxes $10-25, cheese harps $60-80.

Page 51. Butter boxes $18-30, curd rake $45-65, butter spade $50-75, butter packer $25-40, small packer $15-20, longhorn cheese hoop $5-8, tin strainer $10-15, brass strainer with ornate iron handle $35-50, tinned factory dipper $10-18.

Page 52. Process cheese boxes, two-pound size $4-8, five-pound size $10-18.

Page 53. Dairy pans $8-15, large butter box $18-25, small dovetailed butter box $20-25.

Page 54. Ice cream dippers, top left to right: $9, $70, $55, $150, $40, $50, $60-75 $40-50, bottom left to right: 2 small aluminum dippers $4-9, $50-65, $35-40, $6, $9-12, $4-6, $7-12. Framed sign $75-100, bowl and mixer $25-35.

Page 55. Ice cream molds: orange $45-65, turkey $60-85, pumpkin $40-60, santa $65-90, rabbit $60-80, heart $40-60, rose $35-50, turkey $50-75, cow $60-90, Washington hatchet $60-80.

Page 56. Blizzard freezer $30-45, White Mountain freezer $35-55, Snow Peak and other metal freezers $15-30, cardboard container $10-15, ice cream can $5-12, metal freezer $15-30.

Page 57. Bisque figurine $50-65, small cow figures $3-9, stuffed cow $6-10.

Page 58. Three common cow creamers on left $8-15, sitting spotted cow $35-45, cow at lower right $30-45, tin toy cow $55-90, spoon rest $8-15, cow ornaments $2-8.

Page 59. Cow weather vane, complete $150-250, AAA cow $50-65, horned cow figure $125-150, lightning rod cow $90-120.

Page 60. Individual tin animals $60-90, match stick holder $120-200, tray $70-120, Holstein cow and calf $125-200.

Page 61. Jersey cow and calf $90-165, Guernsey cow and calf with envelope $120-180, two-piece Jersey cow $70-90, Success manure spreader tray $45-55.

Page 62. Swiss bell $125-200 (shown), smaller Swiss bells $45-85, Swedish bronze $65-90, common bell $6-15, labeled bell $15-25, bells with straps and wood clappers $30-50.

Page 63. Elsie doll $30-40, Beulah doll $18-25, game $10-12, ice cream box $5-8 Beauregard mug $15-45, creamer $30-45 glass $12-20, sugar and creamer set $55-120, sign $4-6, comic book $3-5, blotter $4-6. Kraft malted milk container $15-30, Thompson's $20-45, small cans $2-5, ice cream box $2-7, butter boxes $3-5.

Page 64. Elsie toy $20-25, thermometers $15-30, Elsie bank $80, egg timer $18-25, ash tray $8-12, other items $8-15, DeLaval button upper left $20-35, Elsie pins $25-35, other buttons and pins $5-12.

Page 65. Hitchhiker creamer $10-15, Harvestor set $40-50, Golden Guernsey set $20-30, small ceramic sets $3-7, playing cards $2-5, calendars $9-12, DeLaval yearbook $4-8.

Page 66. Postcards $15-25 (photograhs), press printed ones are less, ads and pictures $.50-$2, orginal photographs $15-35.

Page 67. Recipe books $3-15, bumper stickers — free to a $1, cow cards $1-5.

Page 68. Metal farm signs: DeLaval $30-40 tin, $75-100 porcelain, Sunset $30-40, Tri-State $30-40, Farm Bureau $20-25, United States Separator $80-100, Artificial Breeding $30-40, Surge $10-20, small cardboard containers $2-6, Cone milk and cream containers $7-14, cardboard yogurt containers $2-4, ceramic yogurt $9-14, ice cream display pieces $7-12.